# ZEBRA SHARKS

by Julie Murray

**Cody Koala**
An Imprint of Pop!
popbooksonline.com

Hello! My name is **Cody Koala**

This book is filled with videos, puzzles, games, and more! Scan the QR codes* while you read, or visit the website below to make this book pop.

## popbooksonline.com/zebra-shark

*Scanning QR codes requires a web-enabled smart device with a QR code reader app and a camera.

**abdobooks.com**

Published by Pop!, a division of ABDO, PO Box 398166, Minneapolis, Minnesota 55439. Copyright ©2024 by Abdo Consulting Group, Inc. International copyrights reserved in all countries. No part of this book may be reproduced in any form without written permission from the publisher. Cody Koala™ is a trademark and logo of Pop!.

Printed in the United States of America, North Mankato, Minnesota.
052023
082023    THIS BOOK CONTAINS RECYCLED MATERIALS

Cover Photo: Blue Planet Archive
Interior Photos: Shutterstock Images; Getty Images; Blue Planet Archive
Editor: Elizabeth Andrews; Grace Hansen
Series Designer: Victoria Bates

**Library of Congress Control Number: 2022950512**

**Publisher's Cataloging-in-Publication Data**
Names: Murray, Julie, author.
Title: Zebra sharks / by Julie Murray
Description: Minneapolis, Minnesota : Pop!, 2024 | Series: Sharks | Includes online resources and index
Identifiers: ISBN 9781098244279 (lib. bdg.) | ISBN 9781098244972 (ebook)
Subjects: LCSH: Zebra shark--Juvenile literature. | Sharks--Juvenile literature. | Sharks--Behavior--Juvenile literature. | Marine fishes--Behavior--Juvenile literature.
Classification: DDC 598.47--dc23

# Table of Contents

**Chapter 1**
Bottom Dwellers . . . . . . . 4

**Chapter 2**
From Stripes to Spots . . . . 8

**Chapter 3**
Nighttime Hunters . . . . . . 14

**Chapter 4**
Life of a Zebra Shark . . . . . 18

Making Connections . . . . . . . . . . . 22
Glossary . . . . . . . . . . . . . . . . . . . . 23
Index . . . . . . . . . . . . . . . . . . . . . . 24
Online Resources . . . . . . . . . . . . . 24

## Chapter 1

# Bottom Dwellers

Zebra sharks live in the western Pacific Ocean, Indian Ocean, and Red Sea. They are often found near coral reefs in warm, shallow waters.

Watch a video here!

Zebra sharks are **bottom-dwelling** sharks.

They spend their days resting on the ocean floor.

Zebra sharks move slowly. They are not a danger to humans.

Chapter 2

# From Stripes to Spots

Zebra sharks grow to an **average** of seven to eight feet (2.1 to 2.4m) long. Their long tail fin makes up half of their total length! They usually weigh around 55 pounds (25kg).

A young zebra shark is dark brown in color with light stripes. The shark gets its name for these zebra-like markings. As the shark grows, its markings will change!

Adult zebra sharks are often confused with leopard sharks. They look very similar.

Adult zebra sharks have tan bodies and many dark spots.

They have **ridges** down their back. They have a large, round head and small eyes.

## Chapter 3

# Nighttime Hunters

Zebra sharks are most active at night. They spend this time hunting. Zebra sharks have long whisker-like **organs** called barbels. These help them find **prey**.

Their **flexible** bodies allow them to squeeze into small spaces where their prey often hide. Zebra sharks eat snails, crabs, small fish, and other sea animals.

## Chapter 4

# Life of a Zebra Shark

A female zebra shark lays up to four eggs at one time. She attaches them to the ocean floor. The eggs hatch five to six months later. Zebra sharks can live for 25 to 30 years.

**Complete an activity here!**

19

Zebra sharks are in danger of dying out. Their biggest threat is humans. Zebra sharks are caught for their meat, **liver oil**, and fins.

# Making Connections

## Text-to-Self

Zebra sharks spend most of their time on the ocean floor. If you were a shark, is that where you would want to spend your time? Why or why not?

## Text-to-Text

Have you ever read a book about another kind of shark? How is that shark like a zebra shark? How is it different?

## Text-to-World

Zebra sharks got their name for the zebra-like markings they have early in life. Can you think of any other animals that got their name because of how they look?

# Glossary

**average** – the usual amount.

**bottom-dwelling** – living and feeding on or near the bed of a sea, lake, or other body of water.

**flexible** – easily bent without breaking.

**liver oil** – the fatty oil obtained from the liver that is used in medicine.

**organ** – a part of an animal or a plant composed of several kinds of tissues. An organ performs a specific function.

**prey** – an animal that is hunted by other animals for food.

**ridge** – a long, narrow, raised section at the top of something.

# Index

baby zebra sharks, 11, 18
barbels, 14
body, 8, 12–13, 17
coloring, 11–12
eggs, 18
food, 14, 17
habitat, 4, 6–7

head, 13
hunting, 14, 17
lifespan, 18
markings, 11–12
size, 8–9
tail, 8–9
threats, 21

# Online Resources

popbooksonline.com

Thanks for reading this Cody Koala book!

This book is filled with videos, puzzles, games, and more! Scan the QR codes* while you read, or visit the website below to make this book pop.

popbooksonline.com/zebra-shark

*Scanning QR codes requires a web-enabled smart device with a QR code reader app and a camera.